FLORA OF TROPICAL EAST AFRICA

COLCHICACEAE

KIM HOENSELAAR[1]

Perrenial herbs, with a tunicated bulb-like or stoloniferous corm. Stem erect or scandent, simple or branched, sometimes almost absent. Leaves basal or cauline, alternate, sometimes opposite and clustered, simple, entire, sessile, often basally sheathing the stem, the leaf apex sometimes ending in a tendril. Inflorescence an umbel, raceme, or spike, or flowers solitary, with or without bracts. Flowers hypogynous, bisexual, regular, sessile or pedicillate. Perianth segments 6, equal, free or basally connate into a tube, sometimes with a basal nectary. Stamens 6; filaments free or inserted at the base of the perianth segments, filiform, sometimes thickened; anthers basifixed or dorsifixed, versatile, usually dehiscing extrorsely by longitudinal slits, sometimes latrorsely to introrsely. Ovary superior, sessile, syncarpous, 3-locular, ovules many, placentation axile; style 1, 3-branched towards the apex, erect or bent at a right angle from the ovary, or styles 3, free, erect or recurved; stigmas 3. Fruit a capsule, septicidal or loculicidal, 3-valved, coriaceous or fleshy; seeds many, (sub)globose, sometimes with a distinct raphe, sometimes red and fleshy.

Nine genera and about 225 species in temperate and tropical Africa, Europe, Asia, Australia and North America.

This family includes several species which are highly toxic and can cause severe damage to livestock. A number of species are used as ornamentals.

1. Stems almost absent; all leaves originating from the base of the plant; inflorescence bracteate, bracts 1.4–2 cm wide . . 1. **Androcymbium**
   Stem distinct; leaves caulescent, inflorescence ebracteate or bracts 0.1–0.4 cm wide 2
2. Plants scandent and/or climbing, at least some leaf tips with tendrils . . . . 2. **Gloriosa**
   Plants erect, leaf tips without tendrils 3
3. Inflorescence single-flowered 4
   Inflorescence several-flowered 6
4. Leaves at least 5 mm wide; flowers showy, perianth segments at least 15 mm long 5
   Leaves up to 5 mm wide; flowers inconspicuous, perianth segments up to 12 mm long . . . . 3. **Iphigenia**
5. Perianth segments free, strongly reflexed, 29–85 mm long; style bent at right angles to the ovary . . . . 2. **Gloriosa**
   Perianth segments connate into a small tube, perianth campanulate, segments 15–40 mm long; style erect . . . 4. **Littonia**
6. Flowers sessile in an ebracteate spike . . . . 5. **Wurmbea**
   Flowers pedicillate in a bracteate inflorescence 7

[1]Taxonomy is partly based on Nordenstam in Notes Roy. Bot. Gard. Edinb. 36: 211–233 (1978); Nordenstam in Opera Bot. 64: 23–24 & 37–39 (1982) & partial FTEA manuscripts by B. Stedje.

7. Leaves at least 15 mm wide; perianth segments 10–23 mm long, 1.5–3.5 mm wide; styles erect, 6–15 mm long . . . . 6. **Ornithoglossum**
Leaves 1.5–5 mm wide; perianth segments 3.5–12 mm long, 0.5–1 mm wide; styles strongly recurved, up to 1 mm long 3. **Iphigenia**

1. ANDROCYMBIUM

Willd. in Ges. Naturf. Fr. Berlin Mag. 2: 21, t. 2 (1808); Baker in J.L.S. 17 (103): 441–446 (1879) & in F.T.A. 7: 559 (1898); K. Krause in N.B.G.B. 7: 512 (1921); Nordenstam in Kubitzki, Fam. & Gen. Vasc. Pl. 3: 182 (1998)

Plants relatively small, with a tunicated, bulb-like corm. Stem erect, almost absent, simple; lowermost leaf a membranous, sheathing, tubular cataphyll, not protracted into a leaf-blade, followed by a few leaves, which are basally sheathing. Leaves narrowly elliptic to linear. Inflorescence a bracteate umbel; pedicel short, erect. Flowers small and inconspicuous, overtopped by large petal-like bracts; perianth segments free, with narrow connivent claws and spreading narrow blades, concave and nectariferous at the base. Stamens inserted at the base of the perianth segments; filaments filiform, thickened towards the base; anthers narrowly oblong, dorsifixed, slightly versatile, dehiscing latrorsely to introrsely by longitudinal slits. Ovary ellipsoid-ovoid; styles 3, erect, free, filiform, stigmatose at the apex, persistent. Capsule ellipsoid, septicidal, coriaceous; seeds subglobose.

About 30 species in Africa and the Mediterranean region; most are found in South Africa.

Androcymbium striatum *A.Rich.* in Tent Fl. Abyss. 2: 336 (1851); K. Krause in N.B.G.B. 7: 521 (1921); Demissew in Fl. Ethiopia and Eritrea 6: 186 (1997); Maroyi in Kirkia 18(1): 6 (2002). Type: Ethiopia, Semien, Entchetkab, *Schimper* II/1338 (K!, syn.) & Ethiopia, Shire, *Quartin Dillon* s.n. (P, syn.)

Herb up to 30 cm high. Corm oblong-ovoid, 1.5–2.5 cm long, 1–1.8 cm in diameter; cataphyll up to 6 cm long. Leaves narrowly elliptic to linear, 13–31 cm long, 0.5–1 cm wide, apex acute to acuminate. Bracts white to greenish, with dark green or purple to brown vascular bundles, (broadly) elliptic, 2.5–8 cm long, 1.4–2 cm wide, apex acuminate to cuspidate; pedicel up to 2(–3) cm long. Flowers yellow-green; perianth segments elliptic-ovate, 5–10 mm long, 3–4 mm wide, base clawed, apex acuminate. Filaments 6–9 mm long; anthers 1–2 mm long. Ovary 5–9 mm long, 2–4 mm wide; styles 2–5 mm long. Capsule 9–13 mm long, 5–7 mm in diameter. Fig. 1 (page 3).

UGANDA. Karamoja District: Kadam Mt, Apr. 1959, *J. Wilson* 758! & Moroto Mt, May 1959, *J. Wilson* 977! & June 1963, *J. Wilson* 1475!

KENYA. Tranz-Nzoia District: SW Elgon, 11 May 1958, *Symes* 378!; S Nyeri District: 3 km N of Kiganjo, 4 km E of the Nairobi–Nanyuki Road, 12 Apr. 1977, *Hooper & Townsend* 1700!; Masai District: foot of the Mau Escarpment on the Kongei Farm road to Narok, 7 Dec. 1969, *Greenway & Kanuri* 13887!

TANZANIA. Mbulu District: Hanang, 26 Dec. 1929, *Burtt* 2283!; Lushoto District: Mtai–Malindi road, near Kidologwai, May 1953, *Drummond & Hemsley* 2661! & W Usambaras, Shumu Forest, Apr. 1953, *Proctor* 165!

DISTR. **U** 1; **K** 3, 4, 6; **T** 1–4, 7; Ethiopia, Angola, South Africa

HAB. Short grassland, often grazed or fire-swept, on shallow soil over rock, sometimes in bushland; 1500–3400 m

CONSERVATION NOTES. Least concern (LC); widely distributed

SYN. *Androcymbium melanthioides* Willd. var. *striatum* (A.Rich.) Baker in J.L.S. 12: 442 (1879) & in F.T.A. 7: 560 (1898); Polhill in Journ. E. Afr. Nat. Hist. Soc. 24: 4 (1962)

Androcymbium melanthioides sensu Baker in Fl. Cap. 6: 517 (1897), pro parte; sensu Hanid in U.K.W.F.: 672 (1974) & U.K.W.F. ed. 2: 322, t. 149 (1994); Blundell in Wild Fl. E. Africa: 417, t. 139 (1982), *non* Willd.

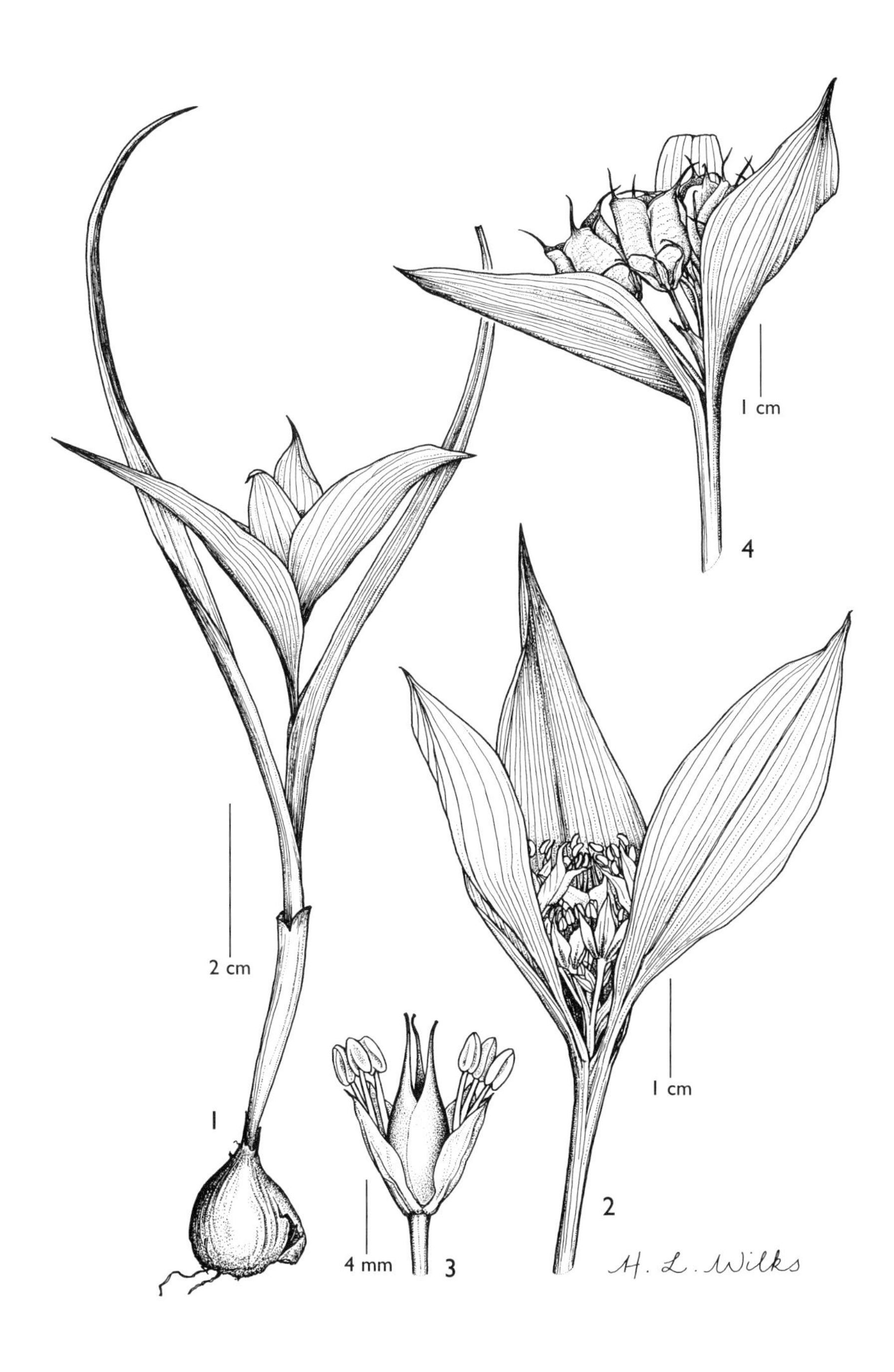

FIG. 1. *ANDROCYMBIUM STRIATUM* — **1**, habit; **2**, inflorescence in detail; **3**, flower; **4**, infructescence. 1–3 from *Greenway & Kanuri* 13887, 4 from *Drummond & Hemsley* 2661. Drawn by H.L. Wilks.

2. GLORIOSA

L. in Sp. Pl. 2 (1753); Baker in J.L.S. 17 (103): 457–458 (1879) & in F.T.A. 7: 563–565 (1898); Field in K.B. 25: 243–244 (1971) & in The genus *Gloriosa*, Lilies and other Liliaceae 1973: 93–95 (1972); Nordenstam in Kubitzki, Fam. & Gen. Vasc. Pl. 3: 183 (1998)

Methonica (Herm.) Juss. in Gen. 48 (1789)
Clinostylis Hochst. in Flora 27:26 (1844)

Plants erect or scandent, simple or branching, with a stoloniferous corm. Leaves alternate, opposite, or verticillate and clustered, sometimes falcate or ending in a tendril. Flowers large and showy, solitary in leaf axils (sometimes corymbose); pedicel erect, recurved apically. Perianth segments free, (strongly) reflexed, clawed at the base, sometimes falcate. Stamens free; filaments filiform or sometimes flattened; anthers narrowly linear-oblong, dorsifixed, versatile, dehiscing latrorsely to extrorsely by longitudinal slits. Ovary (narrowly) elliptic-oblong to oblong; style entire in the lower part, trifid towards the apex with 3 subulate forks obliquely stigmatose at the apex, bent at right angles at the base. Capsule oblong, loculicidal, coriaceous; seeds globose, fleshy, red.

1 species, widespread in Africa and Asia.

Gloriosa superba *L.* in Sp. Pl. 2: 437 (1753); Baker in J.L.S. 17 (103): 457 (1879) & in F.T.A. 7: 563 (1898); Hepper in F.W.T.A. 3 (1): 106 (1968); Field in K.B. 25: 243–244 (1971); Blundell in Wild Fl. E. Africa: 423, t. 409, 488 (1987); Hanid in U.K.W.F. ed. 2: 322, t. 149 (1994); Thulin in Fl. Somalia 4: 67–68 (1995); Demissew in Fl. Ethiopia and Eritrea 6: 184–185 (1997); Maroyi in Kirkia 18(1): 3 (2002). Type: India, Malabaria, *Hermann* 3: 31, no. 122, designated by Wijnands, Bot. Commelins: 133 (1983) (BM, lecto.)

Herb, stem erect, simple or branched, sometimes not higher than 40 cm, or plant scandent or climbing, up to several meters long. Leaves alternate, sometimes opposite or verticillate and clustered, sometimes clustered above the middle of the stem, sessile, base sometimes sheathing the stem, ranging from linear, elliptic-lanceolate, elliptic to ovate, 6–17.5 cm long, 0.4–5 cm wide, apex acute, acuminate, falcate or ending in a tendril. Flowers solitary, axillary, sometimes terminal, different shades of yellow, orange, red, crimson, purple/mauve stripes or fading purple, often bicolored; pedicel erect, recurved apically, 3.5–18 cm long. Perianth sometimes at the base connate into a short tube, up to 2 mm long; perianth segments (strongly) reflexed, base clawed, narrowly elliptic-linear, oblong-lanceolate, sometimes ovate to obovate, 29–85 mm long, 4–25(–38) mm wide, the margins sometimes crisped, apex acuminate to acute, sometimes falcate. Filaments filiform, sometimes flattened, 10–45 mm long; anthers straight to curved, 5.5–15 mm long. Ovary 4–13 mm long, 1–5 mm wide; style 9–50 mm long. Capsule 37–50 mm long, 10–14 mm in diameter; seeds 4 mm.

NOTE. *Gloriosa* undoubtedly exhibits a great variety of forms in the habit, leaf shape and shape and structure of the flower. The differences in habit and floral form have been used to distinguish several "species" in the past based upon specimens collected in Africa. The absence of the leaf tendril and the variation in leaf shape have also been used in the differentiation of species. These characteristics have proven not to be consistent even in individual plants. For example, the presence or absence of leaf tendrils in the erect forms depends entirely on the growth of the plant, and it is frequently only the uppermost leaves of some tall plants which develop tendrils. Plants that develop them one season may do not so in the next. It is not known whether it is possible for an erect, non-climbing plant to go into a climbing habit. Therefore I have decided that the characters used earlier to separate a number of "species" are not adequate for separating taxa, and all material from East Africa is included in *G. superba* L.

Fig. 2. *GLORIOSA SUPERBA* VAR. *SUPERBA* — **1**, habit; **2**, flower; **3**, fruit. 1 and 2 from *Wiltshire* 1, 3 from *Williams* 61. Drawn by H.L. Wilks.

The dwarf growth habit of specimens from Northern Kenya, Ethiopia and Somalia is still recognized as being worthy of taxonomic recognition, but only as a variety. This growth habit seems to be restricted to arid regions.

1. Stem erect and higher than 40 cm, or branched, scandent and climbing; leaves alternate, opposite and somewhat clustered, 1.5–5 cm wide; pedicel 9–18 cm long .... var. *superba*
   Stem erect, plant less than 40 cm high; leaves clustered the above middle of the stem, 0.4–0.8 cm wide; pedicel 3.5–7.5 cm long .... var. *graminifolia*

var. **superba**

Stem erect, simple or branched, plant scandent or climbing, up to several meters long. Leaves alternate, sometimes opposite or verticillate and somewhat clustered, ranging from linear, elliptic-lanceolate, elliptic to ovate, 9–17.5 cm long, (0.8–)1.5–5 cm wide, apex almost always ending in a tendril. Flowers different shades of yellow, orange, red, crimson, purple mauve stripes or fading purple, often bicolored. Pedicel 9–18 cm long. Perianth segments narrowly elliptic-linear, oblong-lanceolate, sometimes ovate to obovate, margins plain, undulate or crisped, 49–85 mm long, 9–25(–38) mm wide. Filaments filiform, sometimes flattened, 11–45 mm long; anthers 7–15 mm long. Ovary 6–13 mm long, 3–5 mm wide; style 20–50 mm long. Capsule 37–50 mm long, 10–14 mm in diameter. Fig. 2 (page 5).

UGANDA. Bunyoro District: Murchison Falls, 24 Sep. 1961, *Rose* 10126!; Ankole District: Mitoma, Igara, Oct. 1938, *Purseglove* 435!; Mengo District; Mabira Forest, 18 Nov. 1938, *Loveridge* 82!

KENYA. Northern Frontier District: Dandu, 03°26'N, 39°94'E, 1 May 1952, *Gillett* 12994!; Trans-Nzoia District: Kitale, near Eldoret, 27 June 1950, *Wiltshire* 1!; Teita District: Sisal estate near Mwatate, 18 Apr. 1960, *Verdcourt & Polhill* 2745!

TANZANIA. Arusha District: Mt Meru, lower slopes, 24 Dec. 1966, *Richards* 21801!; Ufipa District: Mbisi Forest, Malonje Plateau, 17 Mar. 1959, *Richards* 12194!; Chunya District: 30 km W of Chunya, 24 Dec. 1961, *Boaler* 387!

DISTR. **U** 1–4; **K** 1–7; **T** 1–8; **Z**, **P** (fide U.O.P.Z.); widespread in tropical and southern Africa; Madagascar and Asia

HAB. Forest edges, thickets, woodland, short grassland, cultivated land; 0–2100 m

CONSERVATION NOTES. Least concern (LC); widely distributed

SYN. *Gloriosa simplex* L. in Mant. Pl. Alt.: 62 (1767); U.O.P.Z.: 274, fig. (1949); F.P.U.: 203, t. 16 (1962); Verdcourt & Trump, Common Pois. Pl. E.A.: 176, fig. 17 (1969); Field in K.B. 25(2): 243–244 (1971); Cribb & Leedal, Mountain Fl. S. Tanzania: 191, t. 55a (1982). Nomen incerta sedis

Gloriosa virescens Lindl. in Bot. Mag. 52: t. 2539 (1825); Baker in F.T.A. 7: 563 (1898). Type: Mozambique, *Forbes* s.n. (K!, lecto.)

G. abyssinica A.Rich. in Tent. Fl. Abyss. 2: 322 (1851). Type: Ethiopia, Shire Province, *Quartin & Petit* 48 (K!, P, syn.) & Tchélatchérané, *Schimper* s.n. (P, syn.)

Methonica grandiflora Hook. in Bot. Mag. 86: t. 5216 (1860). Type: Equatorial Guinea, Bioko [Fernando Po], *Mann* s.n. (K!, holo.)

Gloriosa virescens Lindl. var. *grandiflora* (Hook.) Baker in J.L.S. 17 (103): 458 (1879). Type: Liberia, Grand Bassa, Niger Exp., *Vogel* 22 (K!, holo.)

G. superba L. var. *angustifolia* Baker in J.L.S. 17 (103): 458 (1879) & in F.T.A. 7: 563 (1898). Type: Mozambique, Lower Rovuma River, *Meller* s.n. (K!, holo.)

G. carsonii Baker in K.B. 1895: 74 (1895). Type: Zambia, Lake Tanganyika, Fwambo, 1894, *Carson* 53 (K!, holo.)

NOTE. All parts of the plant, but especially the corm, are very poisonous; however these plants are also used medicinally to cure e.g. snake bites. Furthermore they are widely used as an ornamental.

In U.O.P.Z. *Gloriosa superba* L. is described as being annual. This is incorrect; it is a perennial plant, each year shooting new stems from its underground corm.

var. **graminifolia** (*Franch.*) *Hoenselaar*, **comb. nov.** Type: Somalia, Karoma Peak, *Révoil* s.n. (P, holo.)

Stem erect, less than 40 cm high; sometimes a cataphyll present, up to 4 cm long. Leaves verticillate and clustered above the middle of the stem, linear, 6–10 cm long, 0.4–0.8 cm wide, apex acute, falcate, sometimes almost ending in a tendril. Flowers often green at the base, all yellow, orange, yellowish red to red; pedicel 3.5–7.5 cm long; perianth segments (narrowly) oblong to lanceolate to obovate, 29–45(–70) mm long, 4–10 mm wide, margins plain. Filaments filiform, up to 10–30 mm long; anthers 5.5–7 mm long. Ovary 4–10 mm long, 1–6 mm wide; style 9–30 mm long. Capsule ± 33 mm long, ± 14 mm in diameter.

KENYA. Turkana District: 17 km S of Lodwar, July 1938, *Pole Evans & Erens* 1574!; locality not known, 610 m, 28 Mar. 1934, *Martin* 96!; Northern Frontier District: between Marsabit and Mt Kulal, near Kargi, Apr. 1986, *Linder* 3604!

DISTR. **K** 1, 2, 4, 7; Ethiopia (S), Somalia

HAB. In open sandy plains, fully exposed, rocky grounds, in open bush land (*Commiphora - Acacia*) and alongside roads; 500–900 m

CONSERVATION NOTES. Least concern (LC); although its distribution is limited, the habitat in which it occurs is fairly common.

SYN. *Gloriosa abyssinica* A.Rich. var. *graminifolia* Franch., Sert. Somal.: 67 (1882).

Littonia baudii Terracc. in Bull. Soc. Bot. Ital. 1892: 425 (1892). Type: Ethiopia, Harerge Region, Ogaden, Gerar Amaden, Apr. 1891, *Baudi & Candeo* s.n. (?FT, holo.)

Gloriosa minor Rendle in J.B. 34: 132 (1896); Baker in F.T.A. 7: 564 (1898); Polhill in Journ. E. Afr. Nat. Hist. Soc. 24: 19 (1962); Blundell in Wild Fl. E. Africa: 423, t. 489 (1987); Hanid in U.K.W.F. ed. 2: 322 (1994). Type: Somalia, W of Shebeli river, 6 Dec. 1894, *Donaldson Smith* s.n. (K!, holo.)

G. baudii (Terracc.) Chiov., Result. Sci. Miss. Stef.-Paoli, Coll. Bot. 1: 176 (1916)

G. graminifolia (Franch.) Chiov., Result. Sci. Miss. Stef.-Paoli, Coll. Bot. 1: 176 (1916)

3. IPHIGENIA

Kunth in Enum. Pl. (Kunth) 4: 212 (1843); Baker in J.L.S. 17 (103): 450–451 (1879) & in F.T.A. 7: 561–562 (1898); Nordenstam in Kubitzki, Fam. & Gen. Vasc. Pl. 3: 182 (1998)

Plants small, with a tunicated, bulb-like corm; stem erect, sometimes flexuous, simple, lowermost leaf or leaves a membranous, sheathing, tubular cataphyll, not protracted into a leaf-blade. Leaves spirally arranged, linear, sessile, basally sheathing the stem. Flowers solitary, axillary in the upper leaves or in a bracteate raceme; bracts linear; pedicel erect or recurved, sometimes elongating in fruit. Flowers small and inconspicuous, perianth segments free, linear to narrowly linear-lanceolate, spreading. Stamens free; filaments filiform, often thickened; anthers oblong, basifixed, versatile, dehiscing extrorsely by longitudinal slits. Ovary oblong; styles 3, free, short, strongly recurved. Capsule oblong to oblong-ovoid, loculicidal, papyraceous; seeds subglobose, papillate, with a distinct raphe.

About 15 species, distributed in Africa, Socotra, India, Australia and New Zealand.

Pedicel erect, 2–2.5 cm long; perianth segments 3.5–6 mm long . . . . 1. *I. pauciflora*
Pedicel recurved, 1.3–2.3 cm long; perianth segments 6–12 mm long . . . . 2. *I. oliveri*

1. **Iphigenia pauciflora** *Martelli* in Fl. Bogos.: 86 (1886); Demissew in Fl. Ethiopia & Eritrea 6: 187 (1997). Type: Eritrea, Keren, *Beccari* 248 (FT, holo.)

Herb up to 35 cm high; corm globose-ovoid, 1–2.2 cm long, 0.6–1.6 cm in diameter; cataphyll when present, 3.1–7 cm long. Leaves linear, 8.5–15 cm long, 0.15–0.5 cm wide, apex acute. Flowers solitary, axillary in the upper leaves or in a bracteate raceme; bracts linear, 2–6 cm long, 0.1–0.25 cm wide; pedicel erect,

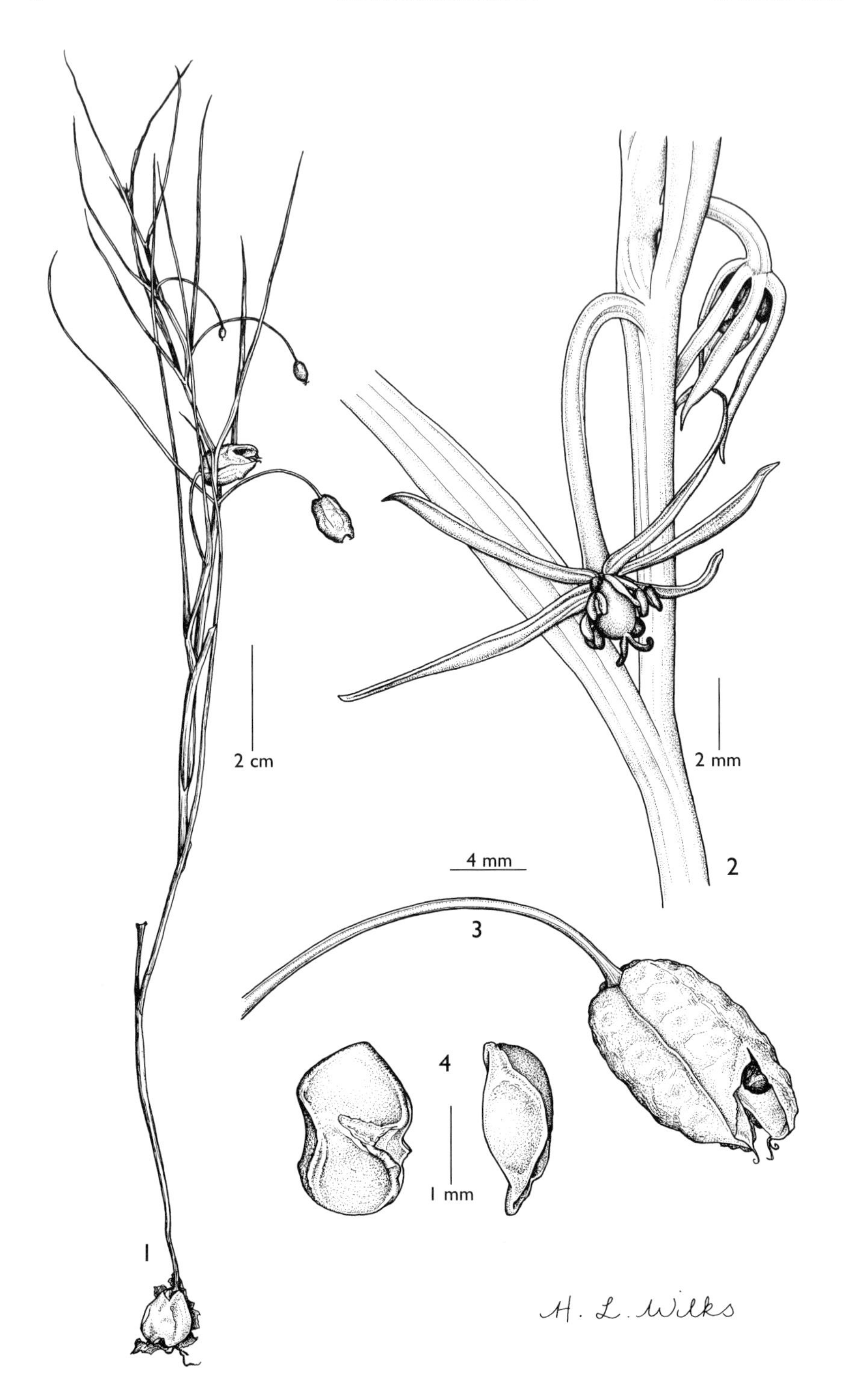

FIG. 3. *IPHIGENIA OLIVERI* — **1**, habit in fruit; **2**, flower; **3**, fruit, **4**, seed, left side view, right front view. 1, 3 and 4 from *Adams* 12, 2 from *Ford* (drawn from picture). Drawn by H.L. Wilks.

2–2.5 cm long. Flowers green to yellow, sometimes purplish on one side, the filaments often dark crimson, dark purplish to almost black, the styles pink to black; perianth segments linear to narrowly linear-lanceolate, 3.5–6 mm long, 0.5–1 mm wide, apex acute. Filaments 1–3.5 mm long; anthers up to 1 mm long. Ovary 2–4 mm long, 1–1.5 mm wide; styles ± 1 mm long. Capsule 0.9–1.9 cm long, 0.4–0.6 cm in diameter; seeds 1–2 mm long, 1–1.2 mm in diameter.

KENYA. Northern Frontier District: Dandu, 14 May 1952, *Gillett* 13059!; Fort Hall District: Thika, N side of Thika River, 11 July 1971, *Faden & Kabuye* 71/555!; Machakos District: 6 km NW of Hunters lodge on Nairobi–Kibwezi Road, Nov. 1975, *Gillett & Brenan* 21700!

TANZANIA. Handeni District: Kwa Mkono, 29 Apr. 1981, *Archbold* 2818!; Iringa District: Msembi–Mbagi, 13 km from Msembi, 17 Dec. 1970, *Greenway & Kanuri* 14807!; Songea District: about 65 km W of Songea, 2 Feb. 1956, *Milne-Redhead & Taylor* 8465!

DISTR. **K** 1, 4; **T** 3, 7, 8; Ivory Coast, Nigeria, Cameroon, Sudan, Eritrea, Ethiopia, Malawi

HAB. Shallow soil over rock, moist hollows in bushland or woodland; 350–1600 m

CONSERVATION NOTES. Least concern (LC); widely distributed

SYN. *Helonia guineensis* Thonn. in Schum. & Thonn., Beskr. Guin. Pl. 182 (1827), *non Iphigenia guineensis* Baker (1879)

Iphigenia ledermannii Engl. & K. Krause in E.J. 45 (1): 127 (1910); Hepper in F.W.T.A. 3 (1): 106 (1968). Type: Cameroon (E), *Ledermann* 4185 & 4323 (B, syn.)

I. abyssinica Chiov. in Ann. Bot. Roma, 9: 147 (1911). Type: Ethiopia, Tigray/Gonder Region, Mai-Aini in Tzelemti, *Chiovenda* 641 (FT, holo.)

I. sudanica A.Chev., Expl. Bot. Afr. Occ. Fr. 1: 658 (1920), nomen

NOTE. *I. pauciflora* Martelli, widespread in tropical Africa, and *I. ledermannii* Engl. & K. Krause, mainly known from West Africa, only seem to differ in flower colour as described respectively in Fl. Ethiopia & Eritrea and F.W.T.A. According to these species descriptions *I. pauciflora* has dark-brown to yellowish-green or greenish-yellow flowers, and the flowers of *I. ledermannii* are deep crimson inside. In the specimens seen in the Flora area, differences other than flower colour could be found, but these differences did not seem to be consistent. The flowers of specimens identified as *I. ledermannii* were not always crimson inside. Variation in flower colour does not seem to be enough to continue separating the taxa; therefore I have decided, for the Flora area, to unite the species under the oldest name, *I. pauciflora* Martelli.

2. **Iphigenia oliveri** *Engl.* in E.J. 15: 467 (1893); Baker in F.T.A. 7: 562 (1898); Hanid in U.K.W.F.: 674 (1974) & in U.K.W.F. ed. 2: 322 (1994); Thulin in Fl. Somalia 4: 68 (1995); Demissew in Fl. Ethiopia & Eritrea 6: 189 (1997). Type: Kenya, Teita District, Taveta, Oct. 1884, *Johnston* s.n. (K!, iso.)

Herb up to 30 cm high; corm globose-ovoid, ± 1.2 cm long, ± 0.7 cm in diameter; cataphyll when present, 2–3.3 cm long. Leaves linear, 6–15.5 cm long, 0.1–0.3 cm wide, apex acute. Flowers solitary, axillary in the upper leaves or in a bracteate raceme; bracts linear, 2.7–4.7 cm long, 0.1–0.15 cm wide; pedicel recurved, 1.3–2.3 cm long. Flowers ranging from yellow to dark maroon, sometimes bicolored; perianth segments linear to narrowly linear-lanceolate, 6–12 mm long, 0.5–1 mm wide, apex acute. Filaments 1–2 mm long, anthers up to 1 mm long. Ovary 2–2.5 mm long, 1–1.5 mm wide; styles up to 1 mm long. Capsule 0.6–2 cm long, 0.4–0.8 cm in diameter; seeds 1–2.2 mm long, 1–1.5 mm in diameter. Fig. 3 (page 8).

KENYA. Kilifi District: Mitangoni on the Kilifi–Kaloleni road, June 1969, *Adams* 12!; Kwale District: near Mtongwe, May 1999, *Luke et al.* 5925! & Ganze–Dida, June 1995, *Luke & Luke* 4379!

TANZANIA. Moshi District: 45 km from Arusha on the Moshi Road, Apr. 1968, *Greenway & Kanuri* 13462!; Mbulu District: Mbulumbul, Block A.G. 5, June 1944, *Greenway* 6936!; Kilosa District: Tembo, Jan. 1931, *Haarer* 2004!

DISTR. **K** 7; **T** 2, 6; Ethiopia, Somalia, Zambia, Mozambique, Zimbabwe, Botswana, South Africa

HAB. In grassland or bushland, in rock crevices; 0–1550 m

CONSERVATION NOTES. Least concern (LC); widely distributed

SYN. *I. somaliensis* Baker in K.B. 105: 228 (1895). Type: Somalia, Wadaba, *Cole & Lort-Phillips* s.n. (K!, holo.)

I. bechuanica Baker, F.T.A. 7: 562 (1898); Obermeyer in Kirkia 1: 84 (1961); Maroyi in Kirkia 18(1): 8 (2002). Type: Botswana, near Kwebe, *Lugard* 81 (K!, holo.)

NOTE. In F.T.A. Baker (1898) treats *Iphigenia oliveri* Engl., as well as a new species *Iphigenia bechuanica* Baker. The two species are distinguished from each other based on a small number of characteristics, the number of flowers on one plant and the shape of the pedicel. During careful examination of the specimens of the Flora area as well as specimens outside the area, these differences did not appear to be consistent. Plants with few flowers, do not have always ascending pedicels, as plants with many flowers do not have always strongly recurved pedicels. Moreover Obermeyer (1961) describes *I. bechuanica* as a plant with 6–10 flowers with recurved pedicels. This leads to confusion on which characters the species are distinguished. In this treatment I have chosen to treat only one species, *Iphigenia oliveri* Engl. as a true species, which involves all variation in characteristics, the number of flowers and the pedicel being strongly recurved or only ascending.

4. **LITTONIA**

Hook. in Bot. Mag. 79: t. 4723 (1853); Baker in J.L.S. 17 (103): 458–459 (1879) & in F.T.A. 7: 566 (1898); Nordenstam in Kubitzki, Fam. & Gen. Vasc. Pl. 3: 183 (1998)

Plants with a stoloniferous corm; stem erect, simple, leafy. Lowermost leaf or leaves a membranous, sheathing, tubular cataphyll, not protracted into a leaf-blade. Leaves ± distichously arranged, the base sheathing the stem. Flowers solitary, born in the axils of many leaves, bright-colored; pedicel recurved. Perianth campanulate, perianth segments forming a small tube, otherwise free, segments nectariferous and obscurely pouch-shaped at the base. Stamens inserted at the base of the perianth segments; filaments filiform; anthers narrowly linear-oblong, dorsifixed, versatile, dehiscing extrorsely by longitudinal slits. Ovary globose; style erect, entire in lower part, 3-forked above, stigmas minute, capitate. Capsule elliptic-oblong, septicidal, coreacious; seeds globose.

About 8 species in Africa, South Africa to Senegal and Arabia.

Perianth at base green, the segments outside orange, vermillion, light red to red, inside yellow to orange; segments longer than 25 mm . . . . 1. *L. lindenii*
Perianth yellowish green; segments no longer than 20 mm . . . . 2. *L. littonioides*

1. **Littonia lindenii** *Baker*, F.T.A. 7: 566 (1898). Type: Tanzania, Kigoma District: Ujiji, *Linden* s.n. (K!, syn.) & Congo (Kinshasa), Lake Mweru, *Descamps* s.n. (BR, syn.)

Herb up to 65 cm high, stem slender; corm globose, 0.5–0.8 cm long, 0.6–0.7 cm in diameter; cataphyll(s) up to 10.5 cm long. Leaves narrowly elliptic-lanceolate to ovate, 5.5–13 cm long, 1.5–2.5 cm wide, apex acute, sometimes acuminate. Flowers axillary; pedicel up to 4.5 cm long; perianth at the base green, segments orange, vermillion, light red to red outside, yellow to orange inside; perianth tube 1–2(–3) mm long, perianth segments narrowly lanceolate-elliptic, 28–40 mm long, 3–6 mm wide. Stamens less than half as long as the perianth; filaments filiform, 4–8 mm long; anthers 4–6 mm long, 1 mm wide. Ovary 5–8 mm long, 2–2.5 mm wide; style 4–8 mm long. Capsule 13 mm long, 4 mm in diameter. Fig. 4 (page 11).

TANZANIA. Mpanda District: Silkcub Highlands, 2 Dec. 1956, *Richards* 7117! & Lake Tanganyika, Ikola, 2 Nov. 1959, *Richards* 11691!; Songea District: about 16 km S of Gumbiro, 27 Jan. 1956, *Milne-Redhead & Taylor* 8294a!

DISTR. **T** 4, 8; Congo (Kinshasa), Zambia, Malawi

HAB. In woodland or grassland; 750–1500 m

FIG. 4. *LITTONIA LINDENII* — **1**, habit; **2**, flower; **3**, stamens and ovary. 1 from *Richards* 7117, 2 and 3 from *Richards* 11691. Drawn by H.L. Wilks.

CONSERVATION NOTES. Least concern (LC); although its distribution is limited, the habitat in which it occurs is fairly common.

2. **Littonia littonioides** (*Baker*) *K.Krause* in E.J. 57: 235 (1921). Type: Angola: Pungo Andongo, *Welwitsch* 1747 (K!, holo.)

Herb with stem up to 60 cm high, slender; cataphyll up to 19 cm long. Leaves narrowly elliptic-lanceolate to ovate, 9–11 cm long, 1.7–1.8 cm wide, apex acute. Flowers axillary, pedicel up to 4 cm long. Perianth yellowish green; perianth tube 1–1.5 mm long; perianth segments narrowly lanceolate-elliptic, 15–17 mm long, 3–4 mm wide. Stamens less than or half as long as the perianth; filaments filiform, 7–8 mm long; anthers 4–5 mm long, 1 mm wide. Style ± 5 mm long.

TANZANIA. Tabora District: Urambo, 1950, *Moors* 13!; Mpanda District: Kapapa Camp, 28 Oct. 1959, *Richards* 11612!
DISTR. **T** 4; Angola, Zambia. Malawi
HAB. Woodland, grassland; 1000–1500 m
CONSERVATION NOTES. Least concern (LC); although its distribution is limited, the habitat in which it occurs is fairly common.

SYN. *Sandersonia littonioides* Baker in Trans. Linn. Soc., Ser. 2, Bot.1: 262 (1878)
Littonia welwitschii Benth. & Hook.f., G.P. 3: 831 (1883); Baker, F.T.A. 7: 566 (1898).

NOTE. In 1998 Nordal & Bingham (K.B. 53: 479) suggested that *Gloriosa* and *Littonia* should be united under one name, in this case the older name *Gloriosa*. Based on only the two *Littonia* species which are distributed in the Flora area, and based on the differences in their vegetative and floral characters in comparison to *Gloriosa*, I have decided to treat the two genera separately.

5. WURMBEA

Thunb., Nov. Gen. Pl. 1:18, t. 1 (1781); Baker in J.L.S. 17 (103): 435–437 (1879) & in F.T.A. 7: 560 (1898); Nordenstam in Notes Roy. Bot. Gard. Edinb. 36(2): 211–233 (1978) & in Kubitzki, Fam. & Gen. Vasc. Pl. 3: 182 (1998)

Plants relatively small, with a tunicated, bulb-like corm; stem erect, lowermost leaf a membranous, sheathing, tubular cataphyll, not protracted into a leaf-blade, followed by a solitary basal leaf and 1–3 cauline leaves, which are basally sheathing the stem. Leaves linear or linear-filiform, semi-terete, somewhat flattened with a concave or channeled adaxial side, or flattened. Inflorescence a spike, 1-many flowered. Flowers stellate to campanulate. Perianth segments basally united in a short to long tube, spreading; nectary a variously shaped, swollen organ in the lower half of the perianth segments. Stamens inserted at the throat of the perianth tube; filaments filiform; anthers oblong or oblong-ellipsoid, minute, dorsifixed, versatile, dehiscing extrorsely by longitudinal slits. Ovary ellipsoid, oblong to ovoid, sometimes triangular; styles 3, erect, free, short, subulate to falcate; stigma minute, persistent in fruit. Capsule oblong, elliptic-oblong, ovoid or obovoid, sometimes slightly triquetrous, septicidal, coriaceous; seeds subglobose, occasionally compressed or slightly angular, with an appendage or minute beak on one side.

About 40 species in Africa (mainly Cape Region) and Australia.

Wurmbea tenuis (*Hook.f.*) *Baker* in J.L.S. 17 (103): 436 (1879) & in F.T.A. 7: 560 (1898); Polhill in Journ. E. Africa Nat. Hist. Soc. 24: 24 (1962); Hepper in F.T.W.A. 3(1): 107 (1968); Hanid in U.K.W.F.: 674 (1974) & in U.K.W.F. ed. 2: 322, t. 149 (1994); Nordenstam in Notes Roy. Bot. Gard. Edinb. 36(2): 221 (1978). Type: Equatorial Guinea, Bioko [Fernando Poo], *Mann* 1454 (K!, holo.)

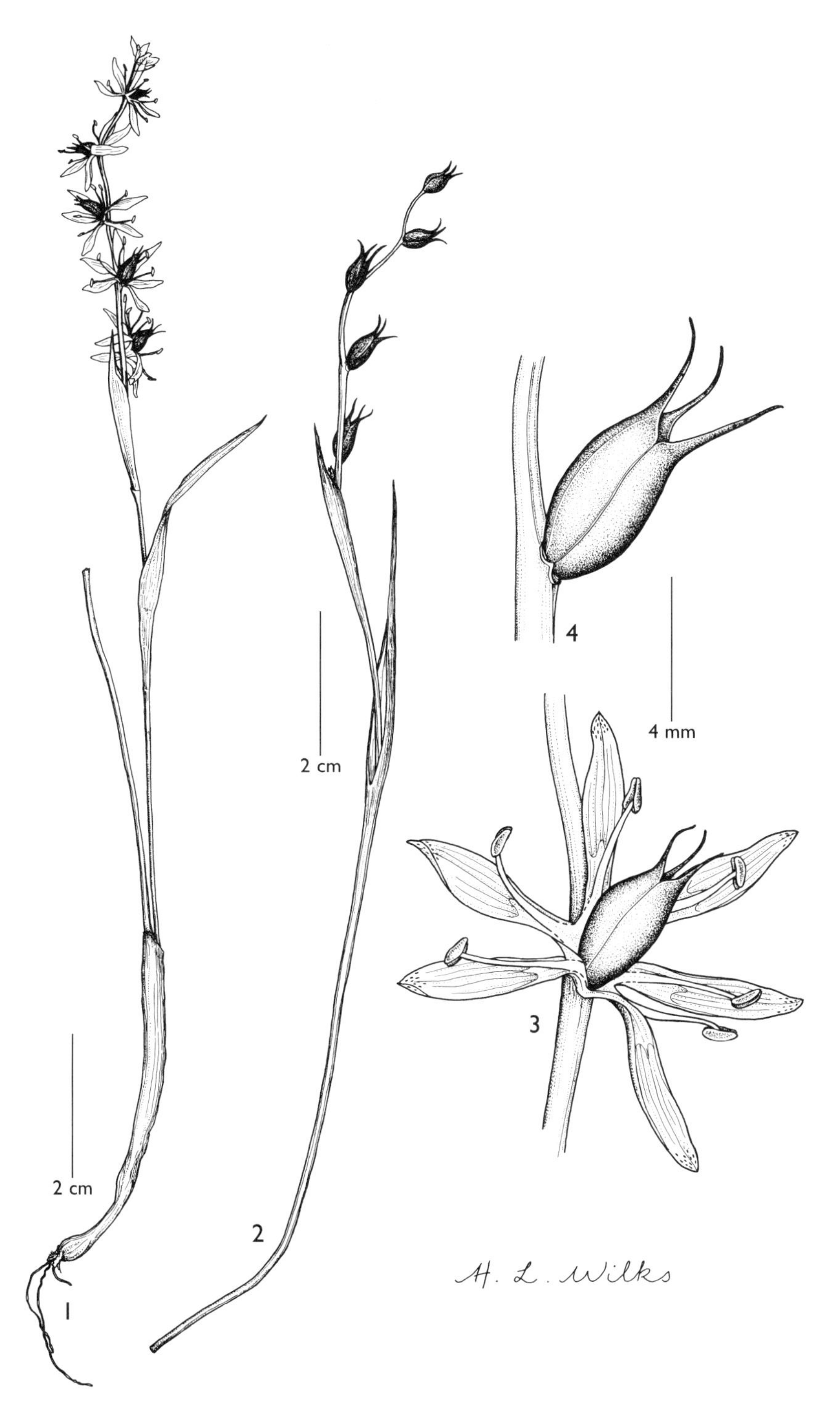

FIG. 5. *WURMBEA TENUIS* SUBSP. *GOETZEI* — **1**, habit in flower; **2**, habit in fruit; **3**, flower; **4**, immature fruit. 1–4 from *Richards* 10321. Drawn by H.L. Wilks.

Herb up to 20 cm high; corm ovoid-subglobose, 9–15 mm long, 7–11 mm in diameter; stem 2–20 cm high, usually slender; cataphyll up to 4.5(–10) cm long. Solitary basal leaf erect, linear or linear-filiform, 3–22(–40) cm long, 0.5 mm wide; cauline leaves erect or recurved, linear, 1–7 cm long, 2–5 mm wide, apex acuminate or cuspidate from a broader sheathing base. Flowers 1–7, sessile; perianth widely campanulate or with stellately spreading segments, 4.5–10 mm long; perianth tube 0.5–1.5 mm long; free segments ovate to lanceolate, 3.5–8 mm long, 0.7–2 mm wide, 3–7 veined, with or without small dots, apically obtuse to subacute. Nectary a longitudinal groove at the base of the free segment, bordered by two ridges apically contiguous or nearly so, closely adjacent to purple blotches below or around the middle of the segment. Filaments 1.5–4 mm long; anthers 0.6–1.5 mm long, 0.3–0.8 mm wide. Ovary 3–4 mm long, 1–2 mm wide; styles 1.5–4 mm long. Capsule 8–15 mm long., 3–5 mm in diameter; seeds 1–1.5 mm in diameter.

SYN. *Melanthium tenue* Hook.f. in J.L.S. 7: 223 (1864). Type as above

NOTE. Within *W. tenuis* Baker, 4 subspecies are recognized, two of which occur in the Flora area. The two other subspecies are subsp. *tenuis*, which occurs in Nigeria, Cameroon and Bioko, and subsp. *australis* B.Nord., which is distributed in South Africa and Lesotho.

Basal leaf linear-filiform, ± semiterete with channeled adaxial side; cauline leaves not overtopping the inflorescence; perianth (6–)7–10 mm long, with stellately spreading segments . . . . subsp. *goetzei*

Basal leaf linear, ± flattened but often with inrolled margins; lower cauline leaf overtopping the inflorescence; perianth 4.5–7.5 mm long, campanulate with erecto-patent or almost spreading segments . . . . subsp. *hamiltonii*

subsp. **goetzei** (*Engl.*) *B.Nord.* in Notes Roy. Bot. Gard. Edinb. 36(2): 225 (1978). Type: Tanzania, Mbeya District, Unyika, Suntas village, *Goetze* 1430 (B, holo., PRE, iso.)

Stem 3–20 cm high. Basal leaf linear-filiform, semi-terete or somewhat flattened with a concave or channeled adaxial side; cauline leaves usually not overtopping the spike. Perianth white to greenish or yellowish cream-coloured or mauve; perianth (6–)7–10 mm long, perianth tube 0.5–1.5 mm long; free segments stellately spreading, marked with a transversely oblong or reniform purple blotch, or two close or contiguous blotches, below the middle of the segment, 5–8 mm long, 1–2 mm wide about the middle with a narrower portion 0.5–1 mm wide, 5–7 veined; apex obtuse-subacute. Filaments 3–4 mm long; anthers 0.8–1.5 mm long, 0.5–0.8 mm wide. Styles filiform-subulate, slender, 2–4 mm long, remaining ± erect or somewhat curving, with slightly elongate stigmas. Fig. 5 (page 13).

TANZANIA. Ufipa District: Mao Village, 14 Dec. 1958, *Richards* 10321!; Mbeya District: Mbosi, 16 Nov.1932, *Davies* 702!; Njombe District: Elton Plateau, 14 Jan. 1961, *Richards* 14148!

DISTR. **T** 4, 7; Congo, Zambia, Malawi

HAB. Grassland, especially in moist sites or where regularly burned, also on thin soil over rock; 1200–2600 m

CONSERVATION NOTES. Least concern (LC); widely distributed

SYN. *W. goetzei* Engl. in E.J. 15: 272 (1893)
W. homblei De Wild. in B.J.B.B. 5: 8 (1915). Type: Congo (Kinshasa), Biano Plateau, near Katentania, *Homblé* 814 (BR, holo., BR, iso).

subsp. **hamiltonii** (*Wendelbo*) *B.Nord.* in Notes Roy. Bot. Gard. Edinb. 36(2): 224 (1978). Type: Kenya, Mt Elgon, track from Endebess, *Wendelbo* 6616 (GB, holo., K!, iso.)

Stem 2–12 cm high. Basal leaf linear, flattened; lower cauline leaf usually overtopping the spike. Perianth white or pale mauve to purple, sometimes white with purple margins, campanulate with ± erecto-patent segments, 4.5–7 mm long, connate for $^{1}/_{5}$–$^{1}/_{3}$; perianth tube

1–1.5 mm long; free segments without small dots or rarely with few oblong minute dots distally, with 1 or 2 deep purple blotches about or below the middle, 3.5–5.5 mm long, 0.8–2 mm wide, 3–7 veined; apex obtuse-rounded. Filaments 2–2.5 mm long, anthers 0.6–0.9 mm long, 0.3–0.6 mm wide. Styles subulate, suberect to somewhat curved, 1.5–2.5 mm long, with subcapitate small stigmas.

UGANDA. Acholi District: Imatong Mts, Langia, Apr. 1943, *Purseglove* 1429!; Karamoja District: Mt Debasien, 30 May 1939, *Thomas* 2940! & Mt Debasien, May 1948, *Eggeling* 5815!
KENYA. Trans-Nzoia District: E Elgon, Suam Saw Mills, May 1941, *Tweedie* 575!; Naivasha District: Kinangop, Apr. 1938, *Chandler* 2415!; N Nyeri District: Sirimon Track, 22 Sep. 1963, *Howard & Verdcourt* 3766!
TANZANIA. Arusha District: Mt Meru, Crater floor, 28 Dec. 1966, *Richards* 21827! & Crater floor, 23 Apr. 1968, *Greenway & Kanuri* 13501!; Kilimanjaro, 13 Jan. 1994, *Grimshaw* 9443!
DISTR. **U** 1, 3; **K** 3, 4, 6; **T** 2; not known elsewhere
HAB. Grassland, especially in thin soil over rock; 2000–3500 m
CONSERVATION NOTES. Least concern (LC); although limited in its distribution, the habitat in which it occurs is fairly common.

SYN. *W. hamiltoni* Wendelbo in Bot. Notis. 121: 114 (1968)

6. ORNITHOGLOSSUM

Salisb., Parad. Lond. 1, Plate 54 (1806); Baker in J.L.S. 17 (103): 448–449 (1879) & in F.T.A. 7: 561 (1898); Nordenstam in Opera Bot. 64: 23 (1982) & in Kubitzki, Fam. & Gen. Vasc. Pl. 3: 183 (1998)

Plants relatively small with a tunicated, bulb-like corm. Stem erect, leafy, of variable length, simple or branched, terminated by a bracteate, racemose inflorescence. Lowermost leaf a membranous, sheathing, tubular cataphyll, not protracted into a leaf-blade. Leaves linear, erecto-patent, canaliculate, the base sheathing the stem. Inflorescence a raceme; bracts linear-lanceolate, simple, upwards successively smaller; pedicel patent, recurved apically. Flowers often in dull shades and often two-colored, sometimes almost black; perianth segments free, persistent, patent or reflexed with a basal tubular or flattened claw; nectary a channel-, pouch-, or pocket-like structure in the claw or in the junction between claw and perianth segment. Stamens distinct, inserted at the base of the perianth segments, filaments curved, filiform or swollen about or below the middle, anthers dorsifixed, versatile, linear-oblong, dehiscing extrorsely by longitudinal slits. Ovary globose or ovoid-oblong, styles 3, free, filiform, curved or spreading, with a small capitate or oblong stigma. Capsule pyriform-globose, loculicidal, coriaceous; seeds many, globose or subglobose, with a distinct raphe.

8 species in Southern and tropical Africa, north to Tanzania.

Ornithoglossum vulgare *B.Nord.* in Opera Bot. 64: 37 (1982); Maroyi in Kirkia 18(1): 7 (2002). Type: Zimbabwe, Gweru District, Gweru Teachers' College, *Biegel* 1467 (SRGH, holo., K!, iso.)

Herb up to 70 cm high; corm oblong-ovoid, up to 4 cm long and 2.5(–4) cm in diameter; stem simple or 2–3, branched, 7–70 cm high; cataphyll up to 7(–14.5) cm long. Leaves glaucous to somewhat greyish green, (2–)4–7(–12), erecto-patent-spreading-recurved, lanceolate, 10–35 cm long, (1–)1.5–5 cm wide near the base, apex acute to subobtuse. Raceme 5–30 cm with up to 25 flowers; bracts linear-lanceolate, 1.9–5.5 cm long, 0.1–0.4 cm wide; pedicels 1.5–7 cm long. Perianth dark brownish, olive-green or sometimes pale lemon-yellow, often with red or purple margins and nectary region; perianth segments lanceolate or linear-lanceolate, (8–)10–23(–25) mm long, 1.5–3.5 mm wide (when expanded), claw (1–)2–3.5(–4) mm

FIG. 6. *ORNITHOGLOSSUM VULGARE* — **1**, habit; **2**, flower; **3** fruit. 1 and 2 from *Brummitt & Polhill* 13667, 3 from *Burtt* 1194. Drawn by H.L. Wilks.

long. Filaments upper parts purplish or mahogany red, lower parts white or light greenish, usually curved, 4–13(–16) mm long; anthers 2.5–6 mm long, often curved. Ovary globose or ovoid-oblong, green; styles spreading, 6–15 mm long. Capsule 1–2.5 cm long, 1–1.5 cm in diameter; seeds 2.5–4 mm in diameter. Fig. 6 (page 16).

TANZANIA. Kondoa District: near Kolo, 5 Jan. 1928, *Burtt* 1194!; Mbeya District: Chimala Scarp, 9 Jan. 1975, *Brummitt & Polhill* 13667!; Masasi District: 57 km W of Maikika and 16 km E of Masasi, 15 Dec. 1955, *Milne-Redhead & Taylor* 7665!

DISTR. **T** 5, 7, 8; Zambia, Malawi, Zimbabwe, Botswana, Namibia, South Africa

HAB. Open *Brachystegia* or *Isoberlinia* woodland; 350–1550 m

CONSERVATION NOTES. Least concern (LC); widely distributed

SYN. *Ornithoglossum glaucum* auctt. mult. e.g. Baker in J.L.S. 17 (103): 449 (1897) & in F.T.A. 7: 561 (1898), *non* Salisb. (1806)

Ornithoglossum viride sensu auctt. e.g. Baker in J.L.S. 17 (103): 449 (1897), *non* (L.f.) Aiton

NOTE. Nordenstam (1982) states in his description of *O. vulgare* that there is a considerable variation within the species, especially in floral details such as tepal size and colour, filament length and anther size. "Some of the extreme forms from the marginal area of the distribution range often give the impression of being distinct taxa, but either intermediates exist connecting them with more typical populations, or the material is too sparse to permit any definite conclusions being made". Some variation between the specimens of the Flora area can also be observed, it is however not consistent and distinct enough to describe separate species, subspecies or varieties for the area.

INDEX TO COLCHICACEAE

New names validated in this part

Gloriosa superba *L.* var. **graminifolia** (*Franch.*) *Hoenselaar* **comb. nov.**

GEOGRAPHICAL DIVISIONS OF THE FLORA

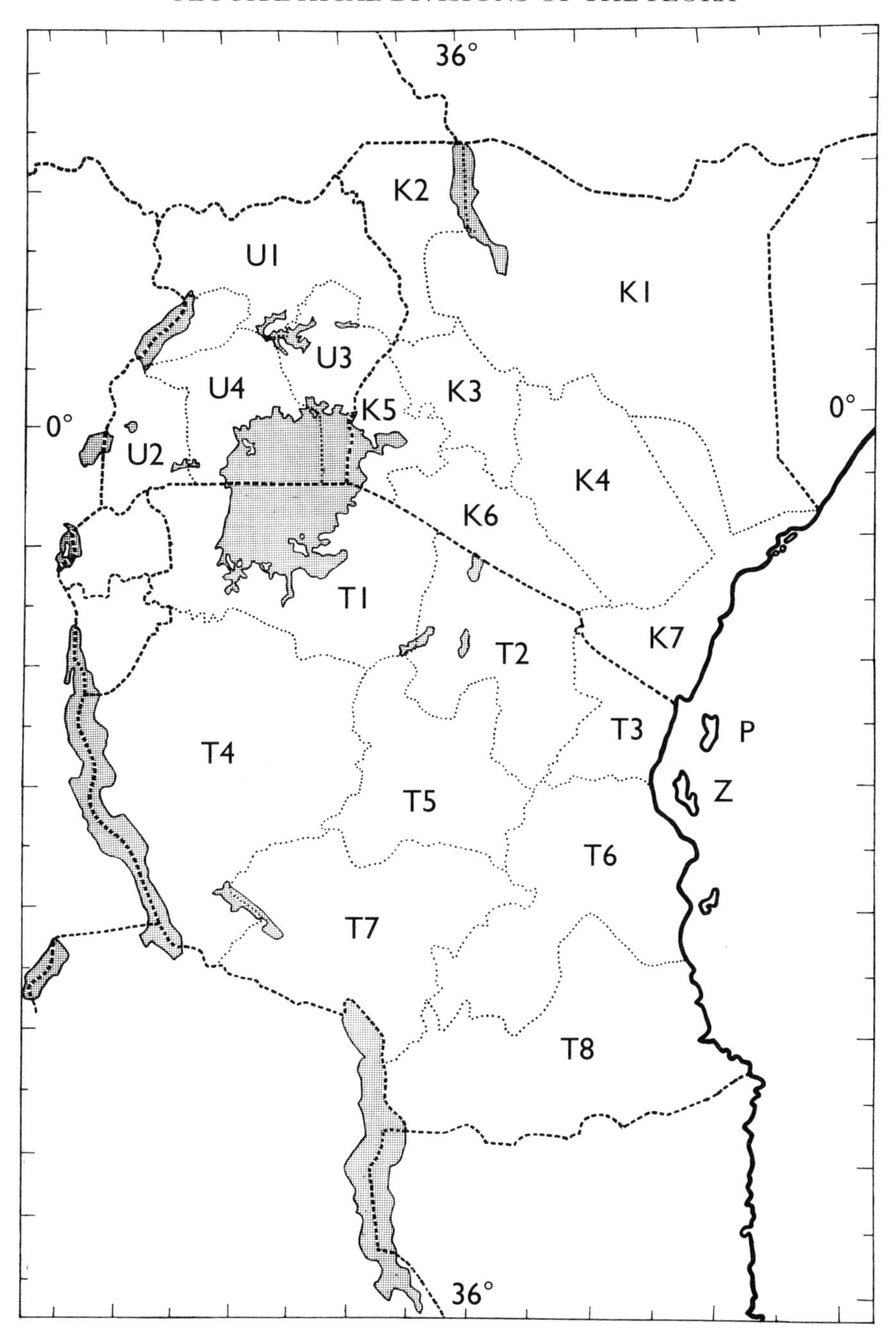

First published in 2005 by
Royal Botanic Gardens, Kew
Richmond, Surrey, TW9 3AB, UK
www.kew.org

ISBN 1 84246 117 6

Design by Media Resources, typesetting and page layout by Margaret Newman,
Information Services Department,
Royal Botanic Gardens, Kew.

Printed by Cromwell Press Ltd.

For information or to purchase all Kew titles please visit
www.kewbooks.com or email publishing@kew.org